On the Study of Zoology

Thomas Henry Huxley

DODO
PRESS

ON THE STUDY OF ZOOLOGY

by Thomas H. Huxley

A Lecture delivered at the South Kensington Museum
in 1861.

ON THE STUDY OF ZOOLOGY*

NATURAL HISTORY is the name familiarly applied to the study of the properties of such natural bodies as minerals, plants, and animals; the sciences which embody the knowledge man has acquired upon these subjects are commonly termed Natural Sciences, in contradistinction to other so-called "physical" sciences; and those who devote themselves especially to the pursuit of such sciences have been and are commonly termed "Naturalists. "

Linnaeus was a naturalist in this wide sense, and his 'Systema Naturae' was a work upon natural history, in the broadest acceptation of the term; in it, that great methodising spirit embodied all that was known in his time of the distinctive characters of minerals, animals, and plants. But the enormous stimulus which Linnaeus gave to the investigation of nature soon rendered it impossible that any one man should write another 'Systema Naturae, ' and extremely difficult for any one to become even a naturalist such as Linnaeus was.

Great as have been the advances made by all the three branches of science, of old included under the title of natural history, there can be no doubt that zoology and botany have grown in an enormously greater ratio than mineralogy; and hence, as I suppose, the name of "natural history" has gradually become more and more definitely attached to these prominent divisions of the subject, and by "naturalist" people have meant more and more distinctly to imply a student of the structure and function of living beings.

However this may be, it is certain that the advance of knowledge has gradually widened the distance between mineralogy and its old associates, while it has drawn zoology and botany closer together; so that of late years it has been found convenient (and indeed necessary) to associate the sciences which deal with vitality and all its phenomena under the common head of "biology"; and the biologists have come to repudiate any blood-relationship with their foster-brothers, the mineralogists.

Certain broad laws have a general application throughout both the animal and the vegetable worlds, but the ground common to these kingdoms of nature is not of very wide extent, and the multiplicity of details is so great, that the student of living beings finds himself obliged to devote his attention exclusively either to the one or the other. If he elects to study plants, under any aspect, we know at once what to call him. He is a botanist, and his science is botany. But if the investigation of animal life be his choice, the name generally applied to him will vary according to the kind of animals he studies, or the particular phenomena of animal life to which he confines his attention. If the study of man is his object, he is called an anatomist, or a physiologist, or an ethnologist; but if he dissects animals, or examines into the mode in which their functions are performed, he is a comparative anatomist or comparative physiologist. If he turns his attention to fossil animals, he is a palaeontologist. If his mind is more particularly directed to the specific description, discrimination, classification, and distribution of animals, he is termed a zoologist.

For the purpose of the present discourse, however, I shall recognise none of these titles save the last, which I shall employ as the equivalent of botanist, and I shall use the term zoology as denoting the whole doctrine of animal life, in contradistinction to botany, which signifies the whole doctrine of vegetable life.

Employed in this sense, zoology, like botany, is divisible into three great but subordinate sciences, morphology, physiology, and distribution, each of which may, to a very great extent, be studied independently of the other.

Zoological morphology is the doctrine of animal form or structure. Anatomy is one of its branches; development is another; while classification is the expression of the relations which different animals bear to one another, in respect of their anatomy and their development.

Zoological distribution is the study of animals in relation to the terrestrial conditions which obtain now, or have obtained at any previous epoch of the earth's history.

Zoological physiology, lastly, is the doctrine of the functions or actions of animals. It regards animal bodies as machines impelled by certain forces, and performing an amount of work which can be expressed in terms of the ordinary forces of nature. The final object of physiology is to deduce the facts of morphology, on the one hand, and those of distribution on the other, from the laws of the molecular forces of matter.

Such is the scope of zoology. But if I were to content myself with the enunciation of these dry definitions, I should ill exemplify that method of teaching this branch of physical science, which it is my chief business to-night to recommend. Let us turn away then from abstract definitions. Let us take some concrete living thing, some animal, the commoner the better, and let us see how the application of common sense and common logic to the obvious facts it presents, inevitably leads us into all these branches of zoological science.

I have before me a lobster. When I examine it, what appears to be the most striking character it presents? Why, I observe that this part which we call the tail of the lobster, is made up of six distinct hard rings and a seventh terminal piece. If I separate one of the middle rings, say the third, I find it carries upon its under surface a pair of limbs or appendages, each of which consists of a stalk and two terminal pieces. So that I can represent a transverse section of the ring and its appendages upon the diagram board in this way.

If I now take the fourth ring, I find it has the same structure, and so have the fifth and the second; so that, in each of these divisions of the tail, I find parts which correspond with one another, a ring and two appendages; and in each appendage a stalk and two end pieces. These corresponding parts are called, in the technical language of anatomy, "homologous parts." The ring of the third division is the "homologue" of the ring of the fifth, the appendage of the former is the homologue of the appendage of the latter. And, as each division exhibits corresponding parts in corresponding places, we say that all the divisions are constructed upon the same plan. But now let us consider the sixth division. It is similar to, and yet different from, the others. The ring is essentially the same as in the other divisions; but the appendages look at first as if they were very different; and yet when we regard them closely, what do we find? A stalk and two terminal divisions, exactly as in the others, but the stalk is very short and very thick, the terminal divisions are very broad and flat, and one of them is divided into two pieces.

I may say, therefore, that the sixth segment is like the others in plan, but that it is modified in its details.

The first segment is like the others, so far as its ring is concerned, and though its appendages differ from any of those yet examined in the simplicity of their structure, parts corresponding with the stem and one of the divisions of the appendages of the other segments can be readily discerned in them.

Thus it appears that the lobster's tail is composed of a series of segments which are fundamentally similar, though each presents peculiar modifications of the plan common to all. But when I turn to the forepart of the body I see, at first, nothing but a great shield-like shell, called technically the "carapace, " ending in front in a sharp spine, on either side of which are the curious compound eyes, set upon the ends of stout movable stalks. Behind these, on the under side of the body, are two pairs of long feelers, or antennae, followed by six pairs of jaws folded against one another over the mouth, and five pairs of legs, the foremost of these being the great pinchers, or claws, of the lobster.

It looks, at first, a little hopeless to attempt to find in this complex mass a series of rings, each with its pair of appendages, such as I have shown you in the abdomen, and yet it is not difficult to demonstrate their existence. Strip off the legs, and you will find that each pair is attached to a very definite segment of the under wall of the body; but these segments, instead of being the lower parts of free rings, as in the tail, are such parts of rings which are all solidly united and bound together; and the like is true of the jaws, the feelers, and the eye-stalks, every pair of which is borne upon its own special segment. Thus the conclusion is gradually forced upon us, that the body of the lobster is composed of as many rings as there are pairs of appendages, namely, twenty in all, but that the six hindmost rings remain free and movable, while the fourteen front rings become firmly soldered together, their backs forming one continuous shield—the carapace.

Unity of plan, diversity in execution, is the lesson taught by the study of the rings of the body, and the same instruction is given still more emphatically by the appendages. If I examine the outermost jaw I find it consists of three distinct portions, an inner, a middle, and an outer, mounted upon a common stem; and if I compare this jaw with the legs behind it, or the jaws in front of it, I find it quite easy to see, that, in the legs, it is the part of the appendage which corresponds with the inner division, which becomes modified into what we know familiarly as the "leg, " while the middle division disappears, and the outer division is hidden under the carapace. Nor is it more difficult to discern that, in the appendages of the tail, the middle division appears again and the outer vanishes; while, on the other hand, in the foremost jaw, the so-called mandible, the inner division only is left; and, in the same way, the parts of the feelers and of the eye-stalks can be identified with those of the legs and jaws.

But whither does all this tend? To the very remarkable conclusion that a unity of plan, of the same kind as that discoverable in the tail or abdomen of the lobster, pervades the whole organization of its skeleton, so that I can return to the diagram representing any one of the rings of the tail, which I drew upon the board, and by adding a third division to each appendage, I can use it as a sort of scheme or plan of any ring of the body. I can give names to all the parts of that figure, and then if I take any segment of the body of the lobster, I can point out to you exactly, what modification the general plan has undergone in that particular segment; what part has remained movable, and what has become fixed to another; what has been excessively developed and metamorphosed and what has been suppressed.

But I imagine I hear the question, How is all this to be tested? No doubt it is a pretty and ingenious way of looking at the structure of any animal; but is it anything more? Does Nature acknowledge, in any deeper way, this unity of plan we seem to trace?

The objection suggested by these questions is a very valid and important one, and morphology was in an unsound state so long as it rested upon the mere perception of the analogies which obtain between fully formed parts. The unchecked ingenuity of speculative anatomists proved itself fully competent to spin any number of contradictory hypotheses out of the same facts, and endless morphological dreams threatened to supplant scientific theory.

Happily, however, there is a criterion of morphological truth, and a sure test of all homologies. Our lobster has not always been what we see it; it was once an egg, a semifluid mass of yolk, not so big as a pin's head, contained in a transparent membrane, and exhibiting not the least trace of any one of those organs, whose multiplicity and complexity, in the adult, are so surprising. After a time a delicate patch of cellular membrane appeared upon one face of this yolk, and that patch was the foundation of the whole creature, the clay out of which it would be moulded. Gradually investing the yolk, it became subdivided by transverse constrictions into segments, the forerunners of the rings of the body. Upon the ventral surface of each of the rings thus sketched out, a pair of bud-like prominences made their appearance—the rudiments of the appendages of the ring. At first, all the appendages were alike, but, as they grew, most of them became distinguished into a stem and two terminal divisions, to which in the middle part of the body, was added a third outer division; and it was only at a later period, that by the modification, or absorption, of certain of these primitive constituents, the limbs acquired their perfect form.

Thus the study of development proves that the doctrine of unity of plan is not merely a fancy, that it is not merely one way of looking at the matter, but that it is the expression of deep-seated natural facts. The legs and jaws of the lobster may not merely be regarded as modifications of a common type, —in fact and in nature they are so, —the leg and the jaw of the young animal being, at first, indistinguishable.

These are wonderful truths, the more so because the zoologist finds them to be of universal application. The investigation of a polype, of a snail, of a fish, of a horse, or of a man, would have led us, though by a less easy path, perhaps, to exactly the same point.

Unity of plan everywhere lies hidden under the mask of diversity of structure—the complex is everywhere evolved out of the simple. Every animal has at first the form of an egg, and every animal and every organic part, in reaching its adult state, passes through conditions common to other animals and other adult parts; and this leads me to another point. I have hitherto spoken as if the lobster were alone in the world, but, as I need hardly remind you, there are myriads of other animal organisms. Of these, some, such as men, horses, birds, fishes, snails, slugs, oysters, corals, and sponges, are not in the least like the lobster. But other animals, though they may differ a good deal from the lobster, are yet either very like it, or are like something that is like it. The cray fish, the rock lobster, and the prawn, and the shrimp, for example, however different, are yet so like lobsters, that a child would group them as of the lobster kind, in contradistinction to snails and slugs; and these last again would form a kind by themselves, in contradistinction to cows, horses, and sheep, the cattle kind.

But this spontaneous grouping into "kinds" is the first essay of the human mind at classification, or the calling by a common name of those things that are alike, and the arranging them in such a manner as best to suggest the sum of their likenesses and unlikenesses to other things.

Those kinds which include no other subdivisions than the sexes, or various breeds, are called, in technical language, species. The English lobster is a species, our cray fish is another, our prawn is another. In other countries, however, there are lobsters, cray fish, and prawns, very like ours, and yet presenting sufficient differences to deserve distinction. Naturalists, therefore, express this resemblance and this diversity by grouping them as distinct species of the same "genus. " But the lobster and the cray fish, though belonging to distinct genera, have many features in common, and hence are grouped together in an assemblage which is called a family. More distant resemblances connect the lobster with the prawn and the crab, which are expressed by putting all these into the same order. Again, more remote, but still very definite, resemblances unite the lobster with the woodlouse, the king crab, the water flea, and the barnacle, and separate them from all other animals; whence they collectively constitute the larger group, or class, 'Crustacea'. But the 'Crustacea' exhibit many peculiar features in common with insects, spiders, and centipedes, so that these are grouped into the still larger assemblage or "province" 'Articulata'; and, finally, the relations which these have to worms and other lower animals, are expressed by combining the whole vast aggregate into the sub-kingdom of 'Annulosa'.

If I had worked my way from a sponge instead of a lobster, I should have found it associated, by like ties, with a great number of other animals into the sub-kingdom 'Protozoa'; if I had selected a fresh-water polype or a coral, the members of what naturalists term the sub-kingdom 'Coelenterata', would have grouped themselves around my type; had a snail been chosen, the inhabitants of all univalve and bivalve, land and water, shells, the lamp shells, the squids, and the sea-mat would have gradually linked themselves on to it as members of the same sub-kingdom of 'Mollusca'; and finally, starting from man, I should have been compelled to admit first, the ape, the rat, the horse, the dog, into the same class; and then the bird, the crocodile, the turtle, the frog, and the fish, into the same sub-kingdom of 'Vertebrata'.

And if I had followed out all these various lines of classification fully, I should discover in the end that there was no animal, either recent or fossil, which did not at once fall into one or other of these sub-kingdoms. In other words, every animal is organized upon one or other of the five, or more, plans, whose existence renders our classification possible. And so definitely and precisely marked is the structure of each animal, that, in the present state of our knowledge, there is not the least evidence to prove that a form, in the slightest degree transitional between any of the two groups 'Vertebrata', 'Annulosa', 'Mollusca', and 'Coelenterata', either exists, or has existed, during that period of the earth's history which is recorded by the geologist. Nevertheless, you must not for a moment suppose, because no such transitional forms are known, that the members of the sub-kingdoms are disconnected from, or independent of, one another. On the contrary, in their earliest condition they are all alike, and the primordial germs of a man, a dog, a bird, a fish, a beetle, a snail, and a polype are, in no essential structural respects, distinguishable.

In this broad sense, it may with truth be said, that all living animals, and all those dead creations which geology reveals, are bound together by an all-pervading unity of organization, of the same character, though not equal in degree, to that which enables us to discern one and the same plan amidst the twenty different segments of a lobster's body. Truly it has been said, that to a clear eye the smallest fact is a window through which the Infinite may be seen.

Turning from these purely morphological considerations, let us now examine into the manner in which the attentive study of the lobster impels us into other lines of research.

Lobsters are found in all the European seas; but on the opposite shores of the Atlantic and in the seas of the southern hemisphere they do not exist. They are, however, represented in these regions by very closely allied, but distinct forms—the 'Homarus Americanus' and the 'Homarus Capensis': so that we may say that the European has one species of 'Homarus'; the American, another; the African, another; and thus the remarkable facts of geographical distribution begin to dawn upon us.

Again, if we examine the contents of the earth's crust, we shall find in the latter of those deposits, which have served as the great burying grounds of past ages, numberless lobster-like animals, but none so similar to our living lobster as to make zoologists sure that they belonged even to the same genus. If we go still further back in time, we discover, in the oldest rocks of all, the remains of animals, constructed on the same general plan as the lobster, and belonging to the same great group of 'Crustacea'; but for the most part totally different from the lobster, and indeed from any other living form of crustacean; and thus we gain a notion of that successive change of the animal population of the globe, in past ages, which is the most striking fact revealed by geology.

Consider, now, where our inquiries have led us. We studied our type morphologically, when we determined its anatomy and its development, and when comparing it, in these respects, with other animals, we made out its place in a system of classification. If we were to examine every animal in a similar manner, we should establish a complete body of zoological morphology.

Again, we investigated the distribution of our type in space and in time, and, if the like had been done with every animal, the sciences of geographical and geological distribution would have attained their limit.

But you will observe one remarkable circumstance, that, up to this point, the question of the life of these organisms has not come under consideration. Morphology and distribution might be studied almost as well, if animals and plants were a peculiar kind of crystals, and possessed none of those functions which distinguish living beings so remarkably. But the facts of morphology and distribution have to be accounted for, and the science, whose aim it is to account for them, is Physiology.

Let us return to our lobster once more. If we watched the creature in its native element, we should see it climbing actively the submerged rocks, among which it delights to live, by means of its strong legs; or swimming by powerful strokes of its great tail, the appendages of whose sixth joint are spread out into a broad fan-like propeller: seize it, and it will show you that its great claws are no mean weapons of offence; suspend a piece of carrion among its haunts, and it will greedily devour it, tearing and crushing the flesh by means of its multitudinous jaws.

Suppose that we had known nothing of the lobster but as an inert mass, an organic crystal, if I may use the phrase, and that we could suddenly see it exerting all these powers, what wonderful new ideas and new questions would arise in our minds! The great new question would be, "How does all this take place? " the chief new idea would be, the idea of adaptation to purpose, —the notion, that the constituents of animal bodies are not mere unconnected parts, but organs working together to an end. Let us consider the tail of the lobster again from this point of view. Morphology has taught us that it is a series of segments composed of homologous parts, which undergo various modifications—beneath and through which a common plan of formation is discernible. But if I look at the same part physiologically, I see that it is a most beautifully constructed organ of locomotion, by means of which the animal can swiftly propel itself either backwards or forwards.

But how is this remarkable propulsive machine made to perform its functions? If I were suddenly to kill one of these animals and to take out all the soft parts, I should find the shell to be perfectly inert, to have no more power of moving itself than is possessed by the machinery of a mill when disconnected from its steam-engine or water-wheel. But if I were to open it, and take out the viscera only, leaving the white flesh, I should perceive that the lobster could bend and extend its tail as well as before. If I were to cut off the tail, I should cease to find any spontaneous motion in it; but on pinching any portion of the flesh, I should observe that it underwent a very curious change—each fibre becoming shorter and thicker. By this act of contraction, as it is termed, the parts to which the ends of the fibre are attached are, of course, approximated; and according to the relations of their points of attachment to the centres of motions of the different rings, the bending or the extension of the tail results. Close observation of the newly-opened lobster would soon show that all its movements are due to the same cause—the shortening and thickening of these fleshy fibres, which are technically called muscles.

Here, then, is a capital fact. The movements of the lobster are due to muscular contractility. But why does a muscle contract at one time and not at another? Why does one whole group of muscles contract when the lobster wishes to extend his tail, and another group when he desires to bend it? What is it originates, directs, and controls the motive power?

Experiment, the great instrument for the ascertainment of truth in physical science, answers this question for us. In the head of the lobster there lies a small mass of that peculiar tissue which is known as nervous substance. Cords of similar matter connect this brain of the lobster, directly or indirectly, with the muscles. Now, if these communicating cords are cut, the brain remaining entire, the power of exerting what we call voluntary motion in the parts below the section is destroyed; and on the other hand, if, the cords remaining entire, the brain mass be destroyed, the same voluntary mobility is equally lost. Whence the inevitable conclusion is, that the power of originating these motions resides in the brain, and is propagated along the nervous cords.

In the higher animals the phenomena which attend this transmission have been investigated, and the exertion of the peculiar energy which resides in the nerves has been found to be accompanied by a disturbance of the electrical state of their molecules.

If we could exactly estimate the signification of this disturbance; if we could obtain the value of a given exertion of nerve force by determining the quantity of electricity, or of heat, of which it is the equivalent; if we could ascertain upon what arrangement, or other condition of the molecules of matter, the manifestation of the nervous and muscular energies depends (and doubtless science will some day or other ascertain these points), physiologists would have attained their ultimate goal in this direction; they would have determined the relation of the motive force of animals to the other forms of force found in nature; and if the same process had been successfully performed for all the operations which are carried on in, and by, the animal frame, physiology would be perfect, and the facts of morphology and distribution would be deducible from the laws which physiologists had established, combined with those determining the condition of the surrounding universe.

There is not a fragment of the organism of this humble animal whose study would not lead us into regions of thought as large as those which I have briefly opened up to you; but what I have been saying, I trust, has not only enabled you to form a conception of the scope and purport of zoology, but has given you an imperfect example of the manner in which, in my opinion, that science, or indeed any physical science, may be best taught. The great matter is, to make teaching real and practical, by fixing the attention of the student on particular facts; but at the same time it should be rendered broad and comprehensive, by constant reference to the generalizations of which all particular facts are illustrations. The lobster has served as a type of the whole animal kingdom, and its anatomy and physiology have illustrated for us some of the greatest truths of biology. The student who has once seen for himself the facts which I have described, has had their relations explained to him, and has clearly comprehended them, has, so far, a knowledge of zoology, which is real and genuine, however limited it may be, and which is worth more than all the mere reading knowledge of the science he could ever acquire. His zoological information is, so far, knowledge and not mere hear-say.

And if it were my business to fit you for the certificate in zoological science granted by this department, I should pursue a course precisely similar in principle to that which I have taken to-night. I should select a fresh-water sponge, a fresh-water polype or a 'Cyanaea', a fresh-water mussel, a lobster, a fowl, as types of the five primary divisions of the animal kingdom. I should explain their structure very fully, and show how each illustrated the great principles of zoology. Having gone very carefully and fully over this ground, I should feel that you had a safe foundation, and I should then take you in the same way, but less minutely, over similarly selected illustrative types of the classes; and then I should direct your attention to the special forms enumerated under the head of types, in this syllabus, and to the other facts there mentioned.

That would, speaking generally, be my plan. But I have undertaken to explain to you the best mode of acquiring and communicating a knowledge of zoology, and you may therefore fairly ask me for a more detailed and precise account of the manner in which I should propose to furnish you with the information I refer to.

My own impression is, that the best model for all kinds of training in physical science is that afforded by the method of teaching anatomy, in use in the medical schools. This method consists of three elements—lectures, demonstrations, and examinations.

The object of lectures is, in the first place, to awaken the attention and excite the enthusiasm of the student; and this, I am sure, may be effected to a far greater extent by the oral discourse and by the personal influence of a respected teacher than in any other way. Secondly, lectures have the double use of guiding the student to the salient points of a subject, and at the same time forcing him to attend to the whole of it, and not merely to that part which takes his fancy. And lastly, lectures afford the student the opportunity of seeking explanations of those difficulties which will, and indeed ought to, arise in the course of his studies.

But for a student to derive the utmost possible value from lectures, several precautions are needful.

I have a strong impression that the better a discourse is, as an oration, the worse it is as a lecture. The flow of the discourse carries you on without proper attention to its sense; you drop a word or a phrase, you lose the exact meaning for a moment, and while you strive to recover yourself, the speaker has passed on to something else.

The practice I have adopted of late years, in lecturing to students, is to condense the substance of the hour's discourse into a few dry propositions, which are read slowly and taken down from dictation; the reading of each being followed by a free commentary expanding and illustrating the proposition, explaining terms, and removing any difficulties that may be attackable in that way, by diagrams made roughly, and seen to grow under the lecturer's hand. In this manner you, at any rate, insure the co-operation of the student to a certain extent. He cannot leave the lecture-room entirely empty if the taking of notes is enforced; and a student must be preternaturally dull and mechanical, if he can take notes and hear them properly explained, and yet learn nothing.

What books shall I read? is a question constantly put by the student to the teacher. My reply usually is, "None: write your notes out carefully and fully; strive to understand them thoroughly; come to me for the explanation of anything you cannot understand; and I would rather you did not distract your mind by reading. " A properly composed course of lectures ought to contain fully as much matter as a student can assimilate in the time occupied by its delivery; and the teacher should always recollect that his business is to feed, and not to cram the intellect. Indeed, I believe that a student who gains from a course of lectures the simple habit of concentrating his attention upon a definitely limited series of facts, until they are thoroughly mastered, has made a step of immeasurable importance.

But, however good lectures may be, and however extensive the course of reading by which they are followed up, they are but accessories to the great instrument of scientific teaching—demonstration. If I insist unweariedly, nay fanatically, upon the importance of physical science as an educational agent, it is because the study of any branch of science, if properly conducted, appears to me to fill up a void left by all other means of education. I have the greatest respect and love for literature; nothing would grieve me more than to see literary training other than a very prominent branch of education: indeed, I wish that real literary discipline were far more attended to than it is; but I cannot shut my eyes to the fact, that there is a vast difference between men who have had a purely literary, and those who have had a sound scientific, training.

Seeking for the cause of this difference, I imagine I can find it in the fact that, in the world of letters, learning and knowledge are one, and books are the source of both; whereas in science, as in life, learning and knowledge are distinct, and the study of things, and not of books, is the source of the latter.

All that literature has to bestow may be obtained by reading and by practical exercise in writing and in speaking; but I do not exaggerate when I say, that none of the best gifts of science are to be won by these means. On the contrary, the great benefit which a scientific education bestows, whether as training or as knowledge, is dependent upon the extent to which the mind of the student is brought into immediate contact with facts—upon the degree to which he learns the habit of appealing directly to Nature, and of acquiring through his senses concrete images of those properties of things, which are, and always will be, but approximatively expressed in human language. Our way of looking at Nature, and of speaking about her, varies from year to year; but a fact once seen, a relation of cause and effect, once demonstratively apprehended, are possessions which neither change nor pass away, but, on the contrary, form fixed centres, about which other truths aggregate by natural affinity.

Therefore, the great business of the scientific teacher is, to imprint the fundamental, irrefragable facts of his science, not only by words upon the mind, but by sensible impressions upon the eye, and ear, and touch of the student, in so complete a manner, that every term used, or law enunciated, should afterwards call up vivid images of the particular structural, or other, facts which furnished the demonstration of the law, or the illustration of the term.

Now this important operation can only be achieved by constant demonstration, which may take place to a certain imperfect extent during a lecture, but which ought also to be carried on independently, and which should be addressed to each individual student, the teacher endeavouring, not so much to show a thing to the learner, as to make him see it for himself.

I am well aware that there are great practical difficulties in the way of effectual zoological demonstrations. The dissection of animals is not altogether pleasant, and requires much time; nor is it easy to secure an adequate supply of the needful specimens. The botanist has here a great advantage; his specimens are easily obtained, are clean and wholesome, and can be dissected in a private house as well as anywhere else; and hence, I believe, the fact, that botany is so much more readily and better taught than its sister science. But, be it difficult or be it easy, if zoological science is to be properly studied, demonstration, and, consequently, dissection, must be had. Without it, no man can have a really sound knowledge of animal organization.

A good deal may be done, however, without actual dissection on the student's part, by demonstration upon specimens and preparations; and in all probability it would not be very difficult, were the demand sufficient, to organize collections of such objects, sufficient for all the purposes of elementary teaching, at a comparatively cheap rate. Even without these, much might be effected, if the zoological collections, which are open to the public, were arranged according to what has been termed the "typical principle"; that is to say, if the specimens exposed to public view were so selected that the public could learn something from them, instead of being, as at present, merely confused by their multiplicity. For example, the grand ornithological gallery at the British Museum contains between two and three thousand species of birds, and sometimes five or six specimens of a species. They are very pretty to look at, and some of the cases are, indeed, splendid; but I will undertake to say, that no man but a professed ornithologist has ever gathered much information from the collection. Certainly, no one of the tens of thousands of the general public who have walked through that gallery ever knew more about the essential peculiarities of birds when he left the gallery than when he entered it. But if, somewhere in that vast hall, there were a few preparations, exemplifying the leading structural peculiarities and the mode of development of a common fowl; if the types of the genera, the leading modifications in the skeleton, in the plumage at various ages, in the mode of nidification, and the like, among birds, were displayed; and if the other specimens were put away in a place where the men of science, to whom they are alone useful, could have free access to them, I can conceive that this collection might become a great instrument of scientific education.

The last implement of the teacher to which I have adverted is examination—a means of education now so thoroughly understood that I need hardly enlarge upon it. I hold that both written and oral examinations are indispensable, and, by requiring the description of specimens, they may be made to supplement demonstration.

Such is the fullest reply the time at my disposal will allow me to give to the question—how may a knowledge of zoology be best acquired and communicated?

But there is a previous question which may be moved, and which, in fact, I know many are inclined to move. It is the question, why should training masters be encouraged to acquire a knowledge of this, or any other branch of physical science? What is the use, it is said, of attempting to make physical science a branch of primary education? Is it not probable that teachers, in pursuing such studies, will be led astray from the acquirement of more important but less attractive knowledge? And, even if they can learn something of science without prejudice to their usefulness, what is the good of their attempting to instil that knowledge into boys whose real business is the acquisition of reading, writing, and arithmetic?

These questions are, and will be, very commonly asked, for they arise from that profound ignorance of the value and true position of physical science, which infests the minds of the most highly educated and intelligent classes of the community. But if I did not feel well assured that they are capable of being easily and satisfactorily answered; that they have been answered over and over again; and that the time will come when men of liberal education will blush to raise such questions, —I should be ashamed of my position here to-night. Without doubt, it is your great and very important function to carry out elementary education; without question, anything that should interfere with the faithful fulfilment of that duty on your part would be a great evil; and if I thought that your acquirement of the elements of physical science, and your communication of those elements to your pupils, involved any sort of interference with your proper duties, I should be the first person to protest against your being encouraged to do anything of the kind.

But is it true that the acquisition of such a knowledge of science as is proposed, and the communication of that knowledge, are calculated to weaken your usefulness? Or may I not rather ask, is it possible for you to discharge your functions properly without these aids?

What is the purpose of primary intellectual education? I apprehend that its first object is to train the young in the use of those tools wherewith men extract knowledge from the ever-shifting succession of phenomena which pass before their eyes; and that its second object is to inform them of the fundamental laws which have been found by experience to govern the course of things, so that they may not be turned out into the world naked, defenceless, and a prey to the events they might control.

A boy is taught to read his own and other languages, in order that he may have access to infinitely wider stores of knowledge than could ever be opened to him by oral intercourse with his fellow men; he learns to write, that his means of communication with the rest of mankind may be indefinitely enlarged, and that he may record and store up the knowledge he acquires. He is taught elementary mathematics, that he may understand all those relations of number and form, upon which the transactions of men, associated in complicated societies, are built, and that he may have some practice in deductive reasoning.

All these operations of reading, writing, and ciphering, are intellectual tools, whose use should, before all things, be learned, and learned thoroughly; so that the youth may be enabled to make his life that which it ought to be, a continual progress in learning and in wisdom.

But, in addition, primary education endeavours to fit a boy out with a certain equipment of positive knowledge. He is taught the great laws of morality; the religion of his sect; so much history and geography as will tell him where the great countries of the world are, what they are, and how they have become what they are.

Without doubt all these are most fitting and excellent things to teach a boy; I should be very sorry to omit any of them from any scheme of primary intellectual education. The system is excellent, so far as it goes.

But if I regard it closely, a curious reflection arises. I suppose that, fifteen hundred years ago, the child of any well-to-do Roman citizen was taught just these same things; reading and writing in his own, and, perhaps, the Greek tongue; the elements of mathematics; and the religion, morality, history, and geography current in his time. Furthermore, I do not think I err in affirming, that, if such a Christian Roman boy, who had finished his education, could be transplanted into one of our public schools, and pass through its course of instruction, he would not meet with a single unfamiliar line of thought; amidst all the new facts he would have to learn, not one would suggest a different mode of regarding the universe from that current in his own time.

And yet surely there is some great difference between the civilization of the fourth century and that of the nineteenth, and still more between the intellectual habits and tone of thought of that day and this?

And what has made this difference? I answer fearlessly—The prodigious development of physical science within the last two centuries.

Modern civilization rests upon physical science; take away her gifts to our own country, and our position among the leading nations of the world is gone to-morrow; for it is physical science only, that makes intelligence and moral energy stronger than brute force.

The whole of modern thought is steeped in science; it has made its way into the works of our best poets, and even the mere man of letters, who affects to ignore and despise science, is unconsciously impregnated with her spirit, and indebted for his best products to her methods. I believe that the greatest intellectual revolution mankind has yet seen is now slowly taking place by her agency. She is teaching the world that the ultimate court of appeal is observation and experiment, and not authority; she is teaching it to estimate the value of evidence; she is creating a firm and living faith in the existence of immutable moral and physical laws, perfect obedience to which is the highest possible aim of an intelligent being.

But of all this your old stereotyped system of education takes no note. Physical science, its methods, its problems, and its difficulties, will meet the poorest boy at every turn, and yet we educate him in such a manner that he shall enter the world as ignorant of the existence of the methods and facts of science as the day he was born. The modern world is full of artillery; and we turn out our children to do battle in it, equipped with the shield and sword of an ancient gladiator.

Posterity will cry shame on us if we do not remedy this deplorable state of things. Nay, if we live twenty years longer, our own consciences will cry shame on us.

It is my firm conviction that the only way to remedy it is, to make the elements of physical science an integral part of primary education. I have endeavoured to show you how that may be done for that branch of science which it is my business to pursue; and I can but add, that I should look upon the day when every schoolmaster throughout this land was a centre of genuine, however rudimentary, scientific knowledge, as an epoch in the history of the country.

But let me entreat you to remember my last words. Addressing myself to you, as teachers, I would say, mere book learning in physical science is a sham and a delusion—what you teach, unless you wish to be impostors, that you must first know; and real knowledge in science means personal acquaintance with the facts, be they few or many. *

[footnote] *It has been suggested to me that these words may be taken to imply a discouragement on my part of any sort of scientific instruction which does not give an acquaintance with the facts at first hand.

But this is not my meaning. The ideal of scientific teaching is, no doubt, a system by which the scholar sees every fact for himself, and the teacher supplies only the explanations. Circumstances, however, do not often allow of the attainment of that ideal, and we must put up with the next best system—one in which the scholar takes a good deal on trust from a teacher, who, knowing the facts by his own knowledge, can describe them with so much vividness as to enable his audience to form competent ideas concerning them. The system which I repudiate is that which allows teachers who have not come into direct contact with the leading facts of a science to pass their second-hand information on. The scientific virus, like vaccine lymph, if passed through too long a succession of organisms, will lose all its effect in protecting the young against the intellectual epidemics to which they are exposed.

Lightning Source UK Ltd.
Milton Keynes UK
UKOW04f1310150913

217217UK00001B/34/P

9 781406 589481